REMEMBER THE DIFFERENCE

WHEN THE SIGNS AND SYMPTOMS OF MILD TO MODERATE HEART FAILURE ARE NOT YET UNDER CONTROL, PRESCRIBE THE DIFFERENCE . . .

BURINEX®A
BUMETANIDE/AMILORIDE
IS MORE EFFECTIVE THAN FRUSEMIDE/AMILORIDE*

A randomised double-blind study established that symptoms of heart failure were significantly less common after treatment with Burinex A than frusemide 40mg/amiloride 5mg*

Reference *Bremner, A.D., et al., Curr. Ther. Res, **50**, Supplement A, 65, 1991.

Further information is available on request from:

Leo Laboratories Limited, Longwick Road, PRINCES RISBOROUGH, Bucks. HP27 9RR

PRESENTED TO

DR. S. WEBB.

WITH THE COMPLIMENTS OF

Leo Laboratories Limited, Longwick Road, PRINCES RISBOROUGH, Bucks. HP27 9RR

® Registered Trademark
March 1993

5083

PLANTS IN CARDIOLOGY

Arthur Hollman, MD, FRCP, FLS
Consulting Cardiologist,
University College Hospital, London

Member of the Committee of Management,
Chelsea Physic Garden, London

Articles reprinted from the *British Heart Journal*

Published by the British Medical Journal
Tavistock Square, London WC1H 9JR

First published 1992
Reprinted 1993
Reprinted 1993 (with amendments)

ISBN 0-7279-0744-1

Typeset and printed in Great Britain by
Latimer Trend & Company Ltd, Plymouth

Contents

Acknowledgements

The author and publishers are grateful to the following:

The British Cardiac Society for its generous support in the production of this book; Jane Dawson, who first prompted the author to write the articles for the *British Heart Journal*; and Dr Dennis Krikler, past Editor of the journal, for his encouragement throughout.

Acknowledgement is also due for permission to reproduce illustrations to: Dioscorides Press (an imprint of Timber Press, Inc.) from *The Healing Forest* (Schultes and Raffauf) © 1990 (page 4), the Linnean Society of London (pages 6, 10, 14, 16, 20, 30, 32 and 34), the Royal Botanic Gardens, Kew (pages 2, 12, 18, 22, 24 and 28), and P. B. Tomlinson (1980) *The Biology of Trees Native to Tropical Florida* (published privately, Petersham, Massachusetts, USA) (page 34).

The articles in this book were originally published in the *British Heart Journal* and the references are as follows: amiodarone, nifedipine, sodium cromoglycate 1991;**65**:57; ryanodine 1991;**65**:71; lignocaine (with R Dahlbom) 1991;**65**:165; aspirin 1991;**65**:193; veratrum alkaloids 1991;**65**:286; methyl xanthine diuretics 1991;**54**:310; ergometrine (ergonovine) 1991;**66**:14; dicoumarol and warfarin 1991;**66**:181; verapamil and morphine 1991;**66**:198; quinine and quinidine 1991;**66**:301; atropine 1991;**66**:367; reserpine and ajmaline 1991;**66**:434; cardiac glycosides 1992;**67**:116; procaine and procainamide 1992;**67**:143; dietary pulmonary hypertension 1992;**67**:235; aconitine and arrhythmias 1992;**67**:315; fagarine 1992;**67**:376.

Introduction

Plants are the origin of important cardiovascular drugs and the articles in this book describe the history of their discovery and development. The organic chemicals that we exploit for medical use are the often poisonous secondary compounds—that is, not involved in metabolism. So it is not entirely clear why they are produced. Some of them act as defences against animal or insect predators, or indeed against other plants.

In many cases a medicinal plant was first used in folk medicine and with some of these remedies the drug is actually contained in the plant and can be used without modification as with digoxin and morphine. But in other instances scientific endeavour has been all important in developing new medicines or new uses from the folk remedy. As examples of new medicines one can cite amiodarone, verapamil and aspirin whilst new uses include atropine, reserpine and ryanodine. There have been important contributions too from other sources. Veterinary medicine led to oral anticoagulants and to an animal model for pulmonary hypertension. Academic chemistry produced lignocaine and it was a patient's own experience that gave us quinidine.

There is no obvious explanation for the distribution of medicinal compounds in the plant kingdom. There are about 300 families of plants and just 22 are the source of all the major pharmaceutical drugs. The family Apocynaceae with a total of 2100 species has three which yield important medicines whilst Compositae with 21 000 species has only two minor ones.

There must be more plant medicines awaiting discovery, but how do we go about finding them? There is no simple answer. These articles show that medicines have come from a wide variety of sources, and although it may not be academically satisfying the fact remains that chance findings and serendipity have led to important discoveries. Plant compounds have inspired chemists to develop medicines which are synthetic, but we should not distrust them because they are not natural. The compounds produced by nature's own laboratory are just as much pure chemical substances as are synthetic molecules. What matters is not their origin but proof of efficacy and toxicity from controlled clinical trials. In searching for new plant medicines we have to take a very broad view and support all kinds of research and enquiry. We need to cast our bread upon the waters.

Amiodarone, nifedipine, sodium cromoglycate

Ammi visnaga is a herbaceous Mediterranean plant belonging to the family Umbelliferae, named after *umbra* a shadow or shade. The binomial name is from antiquity and has no special meaning, while in Egypt the plant is called khella. When dried the stout flower stalks were used as toothpicks hence the early French name, "herbe aux curedents".

People living on the Nile delta often had renal stones because urinary schistosomiasis was common. For many centuries khella, which is common in the delta, was known to relieve renal colic. The seeds (fruits) contain the active principle khellin—first extracted in 1879 and identified at Cairo University in 1932 as a chromone. Khellin works by relaxing the smooth muscle of the ureter.

It was, however, a chance observation in 1945 by G V Anrep, a pupil of both Pavlov and Starling, that led to the development from khellin of modern drugs for asthma and heart disease. Anrep was Professor of Pharmacology in Cairo and his technician who had severe angina got renal colic and treated himself with khella. When the man returned to work Anrep perceptively noticed that he no longer had angina and this stimulated him to investigate the effect of khellin on the heart. Using the heart–lung preparation Anrep measured the coronary blood flow in dogs and showed that khellin was an effective and selective coronary vasolidator. Then he did a clinical trial in patients with angina which gave favourable results. His seminal paper reporting these findings was published in the *British Heart Journal* (1946;**8**:171–7.)

This stimulated research elsewhere; and in Belgium the work of R Charlier and J Broekhuysen, who prepared hundreds of compounds with emphasis on the benzofurane portion of khellin, led in 1961 to the synthesis of amiodarone. The name is derived from *am*, to indicate the presence of an amine function; *iod*, for the iodine moiety; and *arone* from benziodarone, an earlier drug in the ketonic benzofurane group.

F Bossert, working for the Bayer company, decided to use khellin as the starting point for his endeavour to find a coronary vasodilator that worked intravenously as well as orally. After 16 years' work he found a promising dihydropyridine compound and two years and 2000 derivatives later he and Vater produced nifedipine. In Britain Anrep's work catalysed research for

1

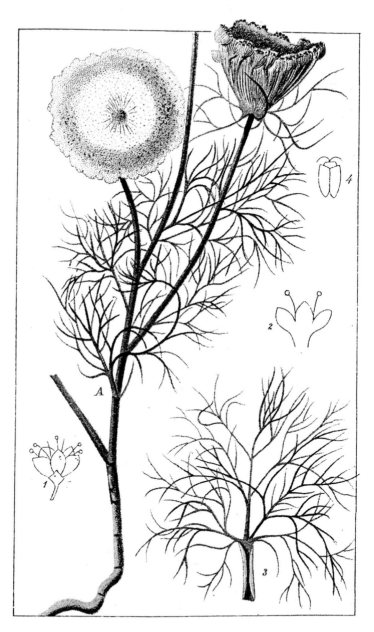

Ammi visnaga (L.) Lam. Bettfreund C. *Flora Argentina*. 1901: plate 155.

a new bronchodilator and sodium cromoglycate, Intal, resulted from the work of an asthmatic doctor, Roger Altounyan, who did all the experiments on himself. Paradoxically its unique mode of action is to prevent the release of mediators of bronchoconstriction and not by bronchodilatation.

Several species of Umbelliferae contain fumocoumarins that are photosensitisers and greatly enhance the action on the skin of ultraviolet radiation. An important one is present in *Ammi majus* L. called ami by Galen and known in Egypt as regl el ghorab. It is a psoralen called methoxsalen and when given orally with long wavelength (ultraviolet A) radiation it constitutes PUVA therapy. This has achieved considerable success in treating psoriasis (long known to improve with sunlight) and also vitiligo, and is useful in mycosis fungoides. Some umbellifers such as *Pastinaca sativa* (wild parsnip) are notorious for producing severe contact dermatitis in sunlight, which reminds one of the photosensitivity of amiodarone. As a genus *Ammi* must be unsurpassed as a source of important medicines.

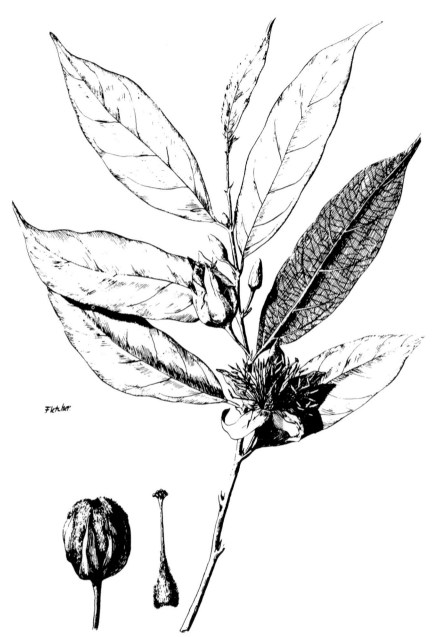

Ryania angustifolia (Turcz.) Monach. Shultes and Raffauf *The Healing Forest.*
1990.

Ryanodine

Ryanodine is a pseudo-alkaloid that is found in *Ryania angustifolia* (Flacourtiaceae) and other species of *Ryania*. It has proved valuable in the study of the calcium release channel of the sarcoplasmic reticulum which it blocks in a dose related manner. Such studies showed that the sarcoplasmic reticulum is a major intracellular calcium store and that it provides most of the energy for cell excitation.

The genus *Ryania* is named after John Ryan, an eighteenth century physician and Fellow of the Royal Society, and its species are small trees in tropical Central and South America. The roots are poisonous and are used to kill rats and alligators, to rid clothing and hair of lice, and even for euthanasia of the elderly by a nomadic Amazonian tribe. Ryanodine was first isolated from *Ryania speciosa* at the Merck Research Laboratories in 1948 during a survey of plant materials for new insecticides.

The family Flacourtiaceae contains a Burmese tree *Hydnocarpus kurzii* whose fruit yields Chaulmoogra oil which was the only effective treatment for leprosy before modern antibacterial drugs. The genus *Idesia* contains salicin, and other genera are used as arrow poisons.

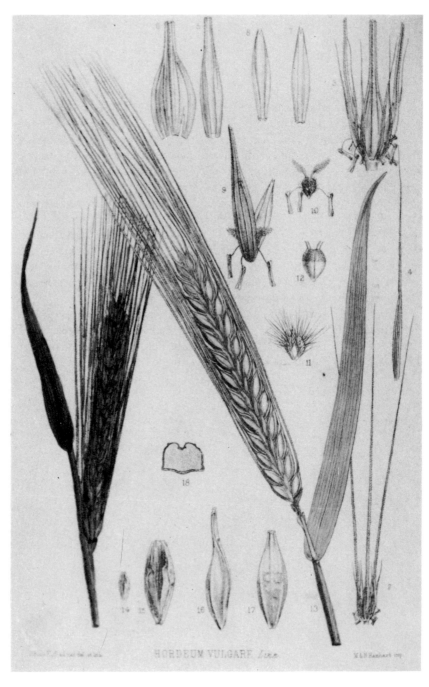

Hordeum vulgare L. Bentley R, Trimen H. *Medicinal plants*. Vol 4. London: Churchill, 1880:293.

Lignocaine

In 1932 Professor Hans von Euler pioneered a new concept when he sought chemical differences between genetically different types of plant. For his first study he chose a chlorophyll deficient mutant of barley, *Hordeum vulgare*, and isolated a compound $C_{11}H_{14}N_2$ which was not present in normal barley, or in other green plants that he tested. He called the compound gramine after the plant's family Gramineae and he thought that it was 2-(dimethylaminomethyl)-indole. But when his assistant Holger Erdtman synthesised this compound in 1935 it proved to be isogramine not gramine, which is 3-(dimethylaminomethyl)-indole. When Erdtman tasted the isogramine he found that it anaesthetised his tongue which gramine did not. But isogramine was too irritant to use as a local anaesthetic and although Erdtman and Nils Lofgren with the support of Astra synthesised several analogues they also could not be used clinically.

The work then stopped for some years but Lofgren later resumed the project, again with Astra's support and in 1943 he produced Xylocaine, which after extensive testing was marketed in 1948. Its generic names are lignocaine and lidocaine.

Gramine was, however, also isolated by Russian workers in 1935 from the great reed *Arundo donax* L. This plant claimed attention because camels refused to eat it because of its bitter taste.

Filipendula ulmaria (L.) Maxim. Meadowsweet or queen of the meadow (*left*); and *Salix alba* L., white willow (*right*). From Gerard J. *The Herball or generall historie of plantes*. 2nd ed. London: Thomas Johnson, 1633.

Aspirin

The success of quinine from 1630 onwards in treating malaria led to its use in other febrile conditions. So when the Reverend Edward Stone in 1763 noticed that powdered bark of the willow tree, *Salix* species (Salicaceae), had a bitter taste like quinine he used it as a substitute for the expensive imported cinchona bark. Its active principle—salicin, which is converted into salicylic acid in the body—was isolated in 1830 and introduced for the treatment of acute rheumatic fever by Dr Thomas Maclagan of Dundee in 1876. He used it in the erroneous belief that the disease was prevalent in cold damp localities and thought that the Doctrine of Signatures indicated that *Salix*, a typical marsh plant, would be nature's remedy. None the less, he got the right answer.

Salicylic acid was originally produced in 1835 from salicylaldehyde found in *Spiraea ulmaria*, now *Filipendula ulmaria* (Rosaceae), the meadowsweet. It became freely available only with the development of a synthetic process in 1874. Its use as an "internal antiseptic" in typhoid fever revealed its antipyretic property. This, together with Maclagan's work, led to its use in rheumatic fever and other rheumatic diseases.

After gastric irritation prevented his father from taking sodium salicylate to treat his chronic arthritis Felix Hoffman, a chemist with the Bayer Company, produced acetylsalicylic acid in 1899 (it had been synthesised elsewhere in 1853). He gave it the trade name Aspirin (*a* for acetyl; *spir* for spiraea; and *in*, a common ending for drugs). This is now the generic name.

The family Salicaceae consists mainly of trees and shrubs from the northern temperate region and has only three genera. The Rosaceae are more numerous (over 100 genera and 3000 species) and more widely distributed. They vary from trees to herbs and include apples, plums, and strawberries. Some of the Rosaceae species yield vitamin C but otherwise neither the Rosaceae nor the Salicaceae contain other important medicines.

Veratrum viride Ait. Bentley R, Trimen H. *Medicinal plants.* vol 4. London: Churchill, 1880:286.

Veratrum alkaloids

The *Veratrum* species (Liliaceae) contain over 20 alkaloids with important cardiovascular, neuromuscular, and respiratory actions. They lower the blood pressure by an unusual mechanism first elucidated in 1867 by A von Bezold. Stimulation of vagal afferent fibres in the left ventricle causes the vasomotor centre to reset homeostatic control at a lower level and to reduce the peripheral vascular resistance through the sympathetic nervous system. When pure alkaloids became available they were shown by A E Doyle and F H Smirk (*British Heart Journal* 1953;15:439–49) and by others to reduce considerably the blood pressure in hypertensive patients. But toxicity, especially nausea and vomiting also caused by vagal action, made treatment difficult and their use stopped despite their attractive vasodilator mode of action. *Veratrum* was used in eclampsia before its pharmacology was studied. It had been given for mania and epilepsy early in the nineteenth century and prompted Dr P de L Baker of Eufala, Alabama, to prescribe *Veratrum viride*, the American or green hellebore, with success in 1859 for a lady with eclamptic fits. This was well before the association between eclampsia and hypertension was recognised. The plant and its alkaloids were used with good effect in toxaemia of pregnancy and eclampsia for the next 100 years.

The *Veratrum* species are handsome perennial plants with large pleated leaves and tall spikes of flowers—well worth growing in the garden. The two European species, *V album* and *V nigrum*, are common on alpine meadows. There are 43 other species, the best known being *V viride* and *V sabadilla*. The alkaloids come from the rhizome and root, or occasionally from the seed. Their common name, from antiquity, is hellebore but they are quite different from the genus *Helleborus*—the Christmas and Lenten roses. *Veratrum* has always been known as a highly poisonous plant causing vomiting, substernal constriction, faintness with a weak pulse, convulsions, and death. In 1985 a Frenchman made wine from it, believing it to be a gentian, and developed complete atrioventricular block. It is also known as a strong teratogen; ewes that eat it can have lambs with a central eye.

Other medicinal members of the Lily family include those that contain cardiac glycosides, such as squill (*Drimia maritima*) and lily of the valley (*Convallaria majalis*). Colchicine is found in the meadow saffron ("autumn crocus") *Colchicum autumnale* while sisal *Agave sisalana* provides the starting material for steroid synthesis.

(*Right*) *Coffea arabica* L. From
Bentley R, Trimen H. *Medicinal plants*.
London: Churchill, 1880:144.

(*Left*) *Camellia sinensis* (L.) Kuntze.
From Curtis W. *Botanical magazine*
1807;**25 and 26**: Tab 998.

(*Right*) *Theobroma cacao* L. From
Kohler FE. *Medizinal-Pflanzen*,
Atlas 1887; Band II: plate 183.

Methyl xanthine diuretics

Caffeine, theobromine, and theophylline were first studied in 1886 and although they are now obsolete they were important before the discovery of the synthetic diuretics in 1951 and especially before organic mercurials were introduced in 1920. They act by depression of renal tubular reabsorption. Caffeine is only a weak diuretic. Theophylline was shown to be better than theobromine by H M Marvin in an excellent early clinical trial (*Journal of the American Medical Association* 1926;**87**:2043–6). It was effective in two thirds of patients who were still oedematous after treatment with digitalis. Theophylline also has a direct stimulant effect on the myocardium which was demonstrated at cardiac catheterisation by McMichael and colleagues at Hammersmith Hospital (*Clinical Science* 1946–48;**6**:125–35) and shown by them to be greater than that of digoxin in hypertensive right heart failure.

Theophylline is found in the tea plant *Camellia sinensis* (Theaceae) a native of China and India but the amount, 0·1%, is small for clinical use so it has to be synthesised. Theobromine is present in the cocoa tree *Theobroma cacao* (Sterculiaceae) which comes from the tropical forests of South America. Caffeine is present in both tea and cocoa and in other traditional beverages. In Africa these are coffee *Coffea arabica* (Rubiaceae) and the cola nut *Sterculia acuminata* (Sterculiaceae). In South America the ancient drink maté is made from a species of holly *Ilex paraguariensis* (Aquifoliaceae) while yoco comes from a species of *Paullinia* (Sapindaceae). Thus the methyl xanthines come from six plant families in the Old and New World that have no obvious botanical similarities. As has so often happened with medicinal plants, their therapeutic value was discovered by chance.

Purgatives were often used to treat heart failure before the xanthine diuretics were introduced and several of them were plants—elaterium, senna, aloes, cascara, croton, rhubarb, and podophyllum. Presumably they produced their effect because of the coincidental sodium loss that accompanied diarrhoea.

Theophylline seems to be making an exciting therapeutic comeback. It is now used to control erythrocytosis in patients with renal transplants. It reduces erythropoietin production by adenosine antagonism (*New England Journal of Medicine* 1990;**323**:86–90).

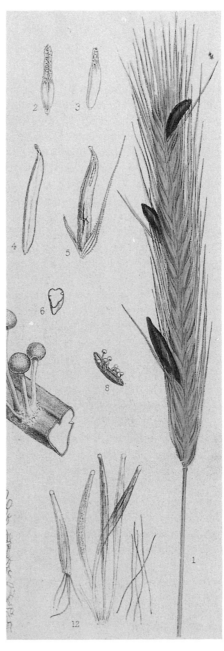

Claviceps purpurea Tul. Bentley R, Trimen H. *Medicinal plants*. vol 4. London: Churchill, 1880:303. The black bodies are the sclerotia.

Ergometrine (ergonovine)

The fungus *Claviceps purpurea* (Clavicipitaceae) is a parasite that infects the flowers of cereals, notably rye, replacing them with a curved hard mass or sclerotium of mycelia which is called ergot (from Old French argot, a cock's spur). In the Middle Ages rye bread infected with ergot caused large epidemics of gangrene of the hands and feet and mental symptoms. Ergotism was often fatal. The blackened extremities looked as if they had been burnt in a fire, and the malady was called St Anthony's Fire because the patients were treated at the saint's monastery in Padua.

Since 300 BC ergot has been recognised as causing abortion. Thus its vasoconstrictor, oxytocic, and cerebral effects were known long before the responsible alkaloids were isolated. Ergotoxine and ergotamine were the first to be found and their effect on human uterine contraction was studied in 1932 by Chassar Moir. He then tested the traditional liquid extract of ergot which was thought by some scientists, though not by clinicians, to be ineffective and he showed its action to be much larger than the two alkaloids. This led to the isolation of a new alkaloid, ergometrine (ergonovine). Its vasoconstrictor action was first used in the investigation of coronary artery spasm in 1976 by T O Cheng at the suggestion of E Shirey and W Sheldon.

Ergot has been called "a veritable treasure house of pharmacological constituents." Ergot derivatives include the hallucinogen, lysergic acid diethylamide (LSD), bromocriptine (used to treat pituitary tumours and parkinsonism), and the serotonin antagonist, methysergide, which is an effective prophylactic in migraine. Methysergide can cause fibrosis of the mitral and aortic valves.

There are over 200 000 species of fungi. They are of great importance in medicine, producing antibiotics and also cyclosporin. The fungi are by tradition classified as plants but because they differ from them and from animals in several ways there is a proposal to create a separate fungal kingdom.

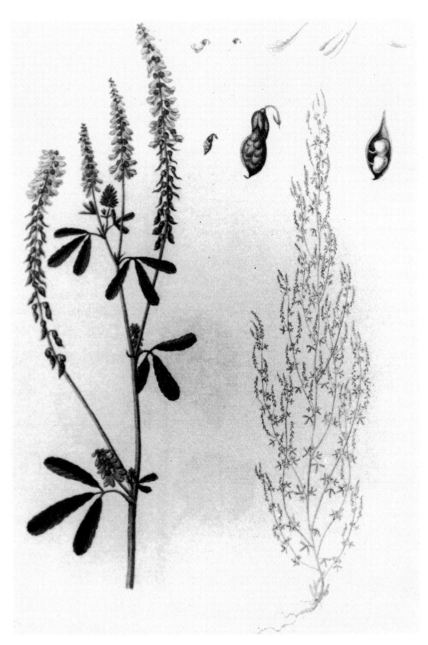

Melilotus officinalis (L.) Lam. *Flora Danicae* 1787: volume vi; plate 934.

Dicoumarol and warfarin

The poor soil, low rainfall, and hard winters of the North American prairies made it difficult to grow crops for animal feed until the melilots or sweet clovers, *Melilotus alba* and *M officinalis* (Leguminosae), were introduced from Europe early this century. They did well and were used for winter feed. In 1922 a new and mysterious disease of cattle was reported in Alberta by a veterinary surgeon F S Schofield who noted that cattle fed on mouldy sweet clover hay were dying of haemorrhage. Properly cured hay was harmless. Schofield found that the clotting time was prolonged: a few years later L M Roderick a veterinary surgeon in Dakota showed that this was due to a reduced crude prothrombin fraction in the blood. The coincidental introduction by Dr A J Quick of his one stage prothrombin method proved essential for further progress. Dr K P Link, who worked in Wisconsin where the disease was common, then took up the search for the enigmatic "haemorrhagic agent". It was six years before the agent was isolated in his laboratory by H A Campbell at dawn on 28 June 1939. It was shown to be a derivative of coumarin—the substance that gives a sweet smell to new mown hay—and was named bishydroxycoumarin. It is formed by fungal action in mouldy sweet clover by oxidation of coumarin to 4-hydroxycoumarin which is then coupled with formaldehyde. On 1 April 1940 it was synthesised. It was first used clinically as an oral anticoagulant at the Mayo Clinic in 1941. The American trade name was dicumarol; and this was adapted in Britain to become the official name dicoumarol.

Link got tuberculosis in 1945 and, having unsuccessfully tried out dicoumarol as a rat poison in 1942, he spent six months in the sanatorium reading about the history of rodent control. From 1946 to 1948 his laboratory staff reappraised the synthetic coumarin derivatives that they had made and found that number 42 had a potent and uniform anticoagulant action. Link proposed it as the ideal rodenticide and coined its name warfarin from the Wisconsin Alumni Research Foundation, which had promoted its use, plus the suffix from coumarin (*Circulation* 1959;**19**:97–107). Though dicoumarol was enthusiastically used by clinicians, warfarin, a rat poison, was ignored—until an army recruit failed to commit suicide after taking a huge dose. Warfarin was soon shown to be better and safer than dicoumarol. It was introduced into clinical practice in 1954.

Papaver somniferum L. Stephenson J, Churchill J M. *Medical botany*. London: 1836, volume 3, plate 159.

Verapamil and morphine

When the unripe seed capsule of the opium poppy *Papaver somniferum* (Papaveraceae) is incised a milky fluid exudes. The dried juice, opium (from the Greek, opos, juice), has been used medicinally for over 5000 years. The 25 or so alkaloids of opium belong to two distinct chemical classes with quite different actions. Morphine and codeine belong to the phenanthrene class. Papaverine, which accounts for only 1% of the alkaloids, is in the benzylisoquinoline class and was isolated in 1848. But its lack of analgesic activity inhibited pharmacological investigation until 1917 when David Macht initiated this at Johns Hopkins Medical School. Because it relaxes smooth muscle it is a good vasodilator; in cardiac muscle it depresses conduction and prevents chloroform induced ventricular fibrillation. But Paul D White found it to be of little use in angina or hypertension and it was tried without success in cardiac arrhythmias. Nevertheless, papaverine and its synthetic analogues were popular as antispasmodic drugs for gastrointestinal and genitourinary ailments and in 1937 the German pharmaceutical firm Knoll asked their chemist, Ferdinand Dengel, to synthesise it. He worked on the compound for over 20 years and in May 1959 he produced an analogue, D365 (D for Dengel), which was soon shown to be pharmacologically much more active than other analogues or other similar drugs. Unlike other vasodilators it had negative inotropic and chronotropic effects. It was iproveratril, later to be called verapamil, and was marketed as Isoptin. Because it was thought to be a β blocker clinical trials in angina began in 1961. Fleckenstein's study of verapamil started in 1963 and led to his seminal discovery of calcium antagonism as its mode of action. In 1972 Schamroth, Krikler, and Garrett (*British Medical Journal* 1972;**i**:660–2) were the first to link the clinical action of verapamil in terminating arrhythmias with Fleckenstein's concept of calcium channel blockade.

Morphine, still pre-eminent for pain relief, was formerly valuable in acute left ventricular failure, and in 1942 Crighton Bramwell and J T King said, "morphine acts as a specific and what is more it is the only drug which is effective."

Papaver somniferum originated in the western Mediterranean and is cultivated chiefly in Asia and Tasmania. Its seed is free of opium and is used on bread. The poppy family, Papaveraceae, has 23 genera and 210 species, mostly in the northern hemisphere. None of its other species has alkaloids that are either better or different from those in the opium poppy.

Cinchona ledgeriana Moens ex Trimen. Bentley R, Trimen H. *Medicinal plants*. Vol 2. London: Churchill, 1880:141.

Quinine and quinidine

The bark of the South American tree *Cinchona* (Rubiaceae) contains quinine, and also quinidine—isolated by Pasteur in 1853. Its use in treating fevers was learnt in Peru by Spanish missionaries who in 1630 brought the bark to Europe where its value in malaria was discovered. In 1749 Jean-Baptiste de Sénac wrote "Long and rebellious palpitations have ceded to this febrifuge". In the nineteenth century quinine was used to augment digitalis therapy, and quinidine was described as "das opium des herzens". But the definitive use of quinidine in arrhythmias came about only because of an astute observation in 1912 by a patient of Professor Karel F Wenckebach who then related the story in the *Journal of the American Medical Association* (1923;**81**:472–4). The patient was a man with attacks of atrial fibrillation who said that "being a Dutch merchant used to good order in his affairs he would like to have good order in his heart business also and asked why there were heart specialists if they could not abolish this very disagreeable phenomenon . . . he knew himself how to get rid of his attacks and as I did not believe him he promised to come back next morning with a regular pulse, and he did".

The man had found by chance that when he took one gram of quinine during an attack it reliably halted it in 25 minutes: otherwise it would last for 2–14 days. Quinine was used then not only in malaria but also as a non-specific remedy for minor ailments as aspirin is today. Wenckebach often tried quinine again but he succeeded in only one other patient. However, it led W Frey in Berlin to study all four cinchona alkaloids in atrial fibrillation and in 1918 he showed that quinidine was the most effective. In 1920 Thomas Lewis put forward his famous hypothesis of circus movement and proposed that quinidine restored normal rhythm by closing the gap between the crest and wake of the circus wave.

The family Rubiaceae is huge with over 10 000 species worldwide. Emetine comes from ipecacuanha and caffeine from coffee but there are no other medicinal species. One genus has the splendid name of *Captaincookia*. Quinine and quinidine are still obtained naturally, from the species *Cinchona ledgeriana* grown commercially in the tropics.

Atropa belladonna L. Stephenson J, Churchill JM. *Medical botany*. London: 1834, volume 1, plate 1.

Atropine

Atropine is obtained from the leaf and berries of *Atropa belladonna* (Solanaceae), the deadly nightshade, an herbaceous plant of central and southern Europe. It is very poisonous and Linnaeus named it after one of the Fates, Atropus, who cut the thread of life. Though it was popular for deliberate poisoning it was avoided medicinally except for external use. The leaf was used to dilate the pupil for cataract extraction, and as a liniment for rheumatism. A large pupil was once held to enhance female beauty, hence belladonna. Nightshade is a curious word which may allude to the narcotic property of the black berries.

Atropine was isolated in 1831, and in 1867 von Bezold showed that it blocked the cardiac effects of vagal stimulation. Sir James Mackenzie (1853–1925) used it widely in his arrhythmia work. He showed that it would revert partial though not complete heart block. He also studied its effect on the rate in digitalised patients with atrial fibrillation and submitted a paper on this subject to *Heart*. The editor, Thomas Lewis, rejected it and this infuriated Mackenzie who replied "You might as well put upon the forefront of the journal 'No articles will be accepted which are not in accordance with the (temporary) beliefs of the Editor'" (McMichael J, *Journal of the Royal College of General Practitioners* 1981;**31**:402–6). Atropine was employed with minimal success to prevent Stokes-Adams attacks and it was little used in cardiology until the introduction of cardiopulmonary resuscitation.

The family Solanaceae is distributed worldwide and has over 2000 species, which include tomato, potato, and tobacco. The invaluable drug hyoscine (scopolamine) is found in *Mandragora officinarum* (mandrake) and in *Datura stramonium* (thornapple). Nowadays atropine and hyoscine are obtained commercially from an Australian tree of this family *Duboisia myoporoides* (corkwood) whose narcotic property was discovered by the aboriginals.

Rauvolfia serpentina Benth. ex Kurz. Rheede tot Draakestein, H A van. *Hortus Indicus Malabaricus*. Amsterdam: Johannes van Someren, 1696, volume 6, fig 47.

Reserpine and ajmaline

Rauvolfia serpentina (Apocynaceae) grows in India where it has been used in Hindu ayurvedic medicine since ancient times. The root resembles a snake and because of the Doctrine of Signatures it was given for snake bite. More importantly the root had a high reputation for calming the insane. In the 1930s psychiatric patients in Calcutta who were being treated with rauwolfia root were found to develop low blood pressure. Later the careful clinical trial of Vakil (*British Heart Journal* 1949;**11**:350–5) established its use in hypertension. Reserpine was the alkaloid that produced this effect and it was isolated in 1951. The term "tranquilliser" was coined in 1953 to describe its then unique calming and sedative effects. Mental depression limited its use in hypertension but another side effect, namely a parkinsonian syndrome, led to a remarkable therapeutic discovery. In 1959 Carlsson found that extrapyramidal symptoms in reserpine treated rabbits were caused by a deficiency of dopamine in the corpus striatum and that its precursor, dopa, reversed the symptoms. He pioneered levodopa treatment for Parkinson's disease.

There are many rauwolfia alkaloids, and in 1931 ajmaline was the first one to be found. It was named after Hakim Azmal Khan who founded the institute in Delhi where it was isolated, but its Class 1 antiarrhythmic action was not discovered until about 1965.

Rauvolfia (spelt with a v) is a genus of the tropical family Apocynaceae with 2100 species, mostly climbers, that are rich in secondary compounds. The vinca alkaloids are found in *Catharanthus roseus*, the Madagascar periwinkle. The cardiac glycosides strophanthus and ouabain occur in *Strophanthus kombe* and *S gratus*.

Nerium oleander L. Gerard J. *The Herball*. London: 1633.

Cardiac glycosides

Foxglove, *Digitalis purpurea* (Scrophulariaceae), was a traditional English folk remedy for dropsy (right heart failure) and its use by an old woman in Shropshire and a carpenter in Oxford led Dr William Withering to undertake his fine scientific study of it from 1775 to 1785. But in spite of his advocacy, its use in dropsy lapsed in the next 100 years though it was used in other diseases such as delirium tremens, epilepsy, and tuberculosis.

By contrast with its neglect in dropsy, its effect on the heart beat gained emphasis. J-B Bouillaud lauded it in 1835 as "le véritable opium du coeur" and gave it by blistering the precordium and covering the area with powdered digitalis. By 1873 F T Roberts had found that digitalis relieved the pulmonary symptoms of mitral disease "especially when there is great irregularity of the heart". However, it was the work of James Mackenzie, a general practitioner in Lancashire that gave digitalis its established position in the treatment of heart failure. His 10 year analysis of the venous pulse in 500 patients led him in 1902 to identify the arrhythmia later shown to be atrial fibrillation and to show that digitalis was of particular value when ". . . heart failure was due to the excessive rate of the ventricle, the ventricle being exhausted for want of rest . . . the effect of the drug was at times phenomenal". He recognised what Withering had not—that the relief of dropsy was directly due to the action on the heart.

A cardiac glycoside was first isolated in 1875 when digitoxin was prepared from *D purpurea*. Soon afterwards ouabain and strophanthin–K were isolated from the African trees *Acokanthera* and *Strophanthus* which were used as arrow poisons. American pioneers learnt from the Indians that *Apocynum*, dogbane, was a diuretic and it became known as the vegetable lancet because it worked as well as venesection. These three genera belong to the family Apocynaceae which has six other members containing glycosides and also includes the genera yielding reserpine and the vinca alkaloids. Altogether in twelve plant families there are 37 genera with glycosides. One of these is the plant used longest as a diuretic, *Drimia maritima* (squill), which was used in Egypt around 1500 BC.

I sometimes wonder whether the eminence of digitalis has inhibited a thorough evaluation of the other glycosides such as the powerful oleandrin found in *Nerium oleander* (*British Heart Journal* 1985;54:258–61).

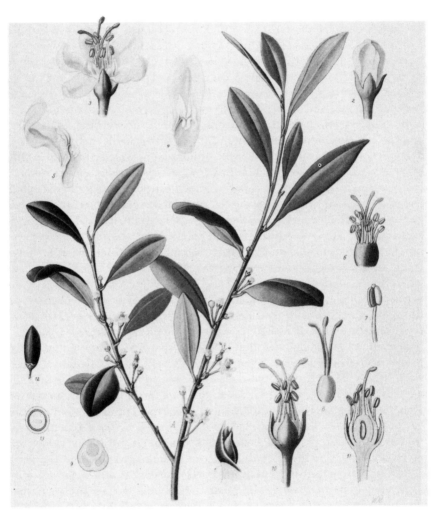

Erythroxylum coca Lam. Kohler F E. *Medizinal-Pflanzen, Atlas.* Gera untermhaus:
E Kohler, 1887, plate 76.

Procaine and procainamide

In the early 1930s Dr Claude S Beck was undertaking pioneer cardiac surgery at the Lakeside Hospital in Cleveland, Ohio. He was attempting to revascularise the heart in angina pectoris by putting a pedicle graft of pectoralis muscle onto the left ventricle and he was also doing pericardiectomy for constrictive pericarditis.

But arrhythmias during and after surgery presented an important problem which was investigated experimentally by Dr Frederick R Mautz. For this study he chose drugs in the cocaine group because they were readily absorbed from mucous membranes and were already known to have some effect on the myocardium. Mautz showed that in dogs procaine produced a monophasic local injury current in the epicardial electrogram and that it prevented extrasystoles when the heart was stimulated electrically (*Journal of Thoracic Surgery* 1936;5:612–28). Procaine had a quinidine-like effect but its action was short-lived owing to esterase action. Its analogue procainamide (Pronestyl) introduced in 1951 had the advantages of being enzyme resistant and active by mouth.

Cocaine was isolated in 1860 from the South American coca plant *Erythroxylum coca* (Erythroxylaceae). This shrub, the "divine plant of the Incas", grows in the eastern Andes and had been used since ancient times to induce a pleasant mental state, to combat fatigue, and increase physical endurance. This folk medicine interested Sigmund Freud and in 1884 he studied the properties of cocaine with the help of his Viennese colleague Carl Koller who was an eye surgeon. When it was found that cocaine numbed the tongue Koller at once realised its potential in ophthalmic surgery. It soon became widely used as the first ever local anaesthetic but its stimulant effect on the nervous system was unwelcome—though Sherlock Holmes took advantage of it. The less toxic synthetic compound procaine was made in 1905.

The small tropical family Erythroxylaceae has no other species with medical uses. It is interesting that two other antiarrhythmic compounds, namely lignocaine and quinine, are also local anaesthetics (pages 7 and 21).

Senecio jacobaea L. *Flora Danicae* Hafnia: Môllerus, 1787: volume 6; plate 944.

Dietary pulmonary hypertension

The idea that certain plants could produce pulmonary hypertension seemed so unlikely that I began to wonder just how this discovery had been made. The story began in Iowa in 1884 when a new disease of horses with hepatic cirrhosis was traced to the ingestion of *Crotalaria sagittalis* (Leguminosae) the native species of the rattlebox plant which was grown as green manure to improve the sandy soil. In 1921 *C spectabilis* was introduced from India and it caused many outbreaks of disease among farm animals in the southern United States. Lesions in cows, pigs, goats, chickens, and horses included subendocardial haemorrhage, thickened pulmonary alveoli, pulmonary oedema, anaemia, and renal and hepatic disease. *Senecio* (Compositae) caused equine cirrhosis in South Africa in 1920, as in North America did *Atalaya* (Sapindaceae), alsike clover *Trifolium* (Leguminosae), and the tar weed *Amsinckia* (Boraginaceae). A recent human cirrhosis epidemic in India followed the accidental ingestion of *Heliotropium* (Boraginaceae). All these genera and families contain pyrrolizidine alkaloids.

But none of these studies of natural or induced disease reported cardiac hypertrophy or pulmonary arterial disease. However, in 1955 Schoental and Head produced pulmonary infarction in rats with *Crotalaria*. Then in 1961 the breakthrough came when J J Lalich in Madison showed that rats fed *Crotalaria* seed or its alkaloid monocrotaline developed acute pulmonary arteritis. He went on to pioneer the crucial long-term study. The rats developed intimal and muscular thickening of the pulmonary arterioles, dilated pulmonary arteries and right ventricular hypertrophy. In 1967 Kay, Harris, and Heath in Birmingham were the first to measure the right heart pressure in treated rats and confirm pulmonary hypertension. Later it was shown that the British plant ragwort, *Senecio jacobaea*, sold in health stores for coughs and colds, produced pulmonary hypertension in rats. Meanwhile Bras and his colleague had discovered that cirrhosis in Jamaican children was caused by hepatic veno-occlusive disease and they showed that their patients' histology was identical with that of animals with *Crotalaria* and *Senecio* poisoning. Jamaican children often drank "bush-tea" made from these plants. But pulmonary hypertension has never been found in patients with veno-occlusive cirrhosis. Maybe only rats are susceptible.

Again, as with sweet clover disease of cattle (page 17), a leguminous plant introduced from abroad for farming purposes has led via veterinary medicine to an important cardiovascular discovery.

31

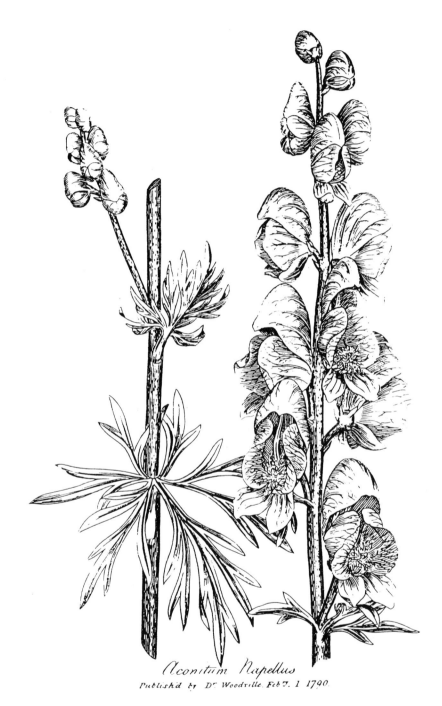

Aconitum Napellus

Publish'd by Dr Woodville. Feby. 1 1790.

Aconitum napellus L. Woodville W. *Medical Botany*. London: Woodville 1790;
volume 1: page 16.

Aconitine and arrhythmias

Various species of aconite, *Aconitum* (Ranunculaceae), such as wolfsbane and monkshood, have been known as deadly poisons since the time of ancient Greece (Medea used aconite to kill Theseus). Soon after the leaf or root is eaten a characteristic tingling spreads over the body, the voluntary muscles twitch, the pulse becomes irregular and weak, and death inexorably follows in a few hours. Animal studies in the nineteenth century showed that the alkaloid aconitine produced vagal slowing followed by an irregular rhythm; and in 1897 S A Matthews at Ann Arbor, using the myocardiograph, became the first to show that it causes atrial and ventricular fibrillation (*Journal of Experimental Medicine* 1897;**2**:593–606). In 1909 in the first paper of the new journal *Heart* A R Cushny showed that it caused pulsus alternans in dogs.

Despite its reputation aconite became a popular medicine and was used to treat neuralgia, fever, pericarditis, and nervous palpitation. In 1869 Sydney Ringer wrote "Perhaps no drug is more valuable than aconite." But in 1880 a patient became very ill when the source of his tincture was unknowingly changed to another aconite root, and his physician Dr Meyer died after taking a dose to justify the safety of his prescription.

It was the work of David Scherf that established the special place of aconitine in experimental arrhythmias. In 1947 he showed that when aconitine was injected into the sinus node of dogs it was better than faradisation in producing episodes of atrial fibrillation or flutter that were long enough to assess the effect of antiarrhythmic drugs. Myron Prinzmetal used topical aconitine for his high-speed cinematograph studies of atrial arrhythmias in 1952.

The family Ranunculaceae has 58 genera and 1750 species mostly in temperate regions. Many of them are poisonous because they contain benzyl isoquinoline and other alkaloids, and some species of *Ranunculus* (buttercup) cause photodermatitis. The genera *Adonis* (pheasant's eye) and *Helleborus* (Christmas rose) contain cardiac glycosides but they are not used medicinally and neither are any other species.

It could still be rewarding to study the cellular mechanism of aconitine induced arrhythmias.

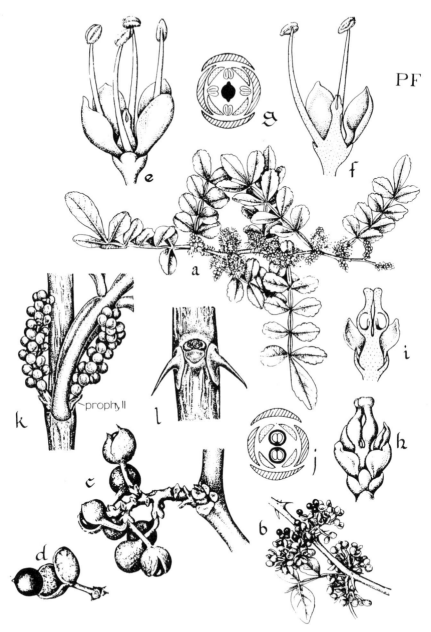

PF

prophyll

Zanthoxylum fagara Tomlinson P. B. *The Biology of Trees Native to Tropical Florida*. 1980.

Fagarine

In 1932 G Stuckert was the first to isolate a new alkaloid, fagarine, from the Argentinian plant *Fagara coco* (Rutaceae); and with A Sartori he showed that it had a depressant action on the myocardium of rabbits. Further work at the University of Cordoba by Moisset de Espanés and others showed that fagarine raised the threshold for atrial and ventricular fibrillation in response to faradic stimulation, and that it decreased the incidence of ventricular fibrillation after coronary ligation in dogs. In all these experiments it was more effective than quinidine. Then A Taquini tried its effect in six patients with atrial flutter or fibrillation who were resistant to quinidine. In all of them intramuscular fagarine restored sinus rhythm within 30 minutes (*Science* 1945;**102**:69–70). In 1948 David Scherf showed that fagarine reliably reverted atrial fibrillation induced by aconitine in dogs (*Proceedings of the Society for Experimental Biology and Medicine* 1948;**67**:59–60).

The genus *Fagara* has been merged with *Zanthoxylum* and the name of the original *F. coco* (Gill.) Engl. is now *Z. coco* Gill ex Hook and Arn. The illustration is of a closely related species. The South African plant *Z. capensis*, a "fever tree", is used medicinally.

The family Rutaceae is widespread, especially in the tropics, and it includes *Pilocarpus microphyllus*, the source of pilocarpine. Citrus fruits belong to this family. Rutaceae is the fourth family of plants described in this book with antiarrhythmic properties—quinidine, procaine, and lignocaine all being derived from other families. It would be interesting to know whether fagarine is still under investigation.

Glossary of selected terms

Ajmaline	An anti arrhythmic (q.v.) drug (Class 1)
Alkaloid	A basic nitrogen containing organic compound found in plants, its name always ends in -ine, e.g. caffeine
Amiodarone	An anti arrhythmic (q.v.) drug (Class 111) with a new mode of action
Analgesic	A pain relieving drug
Angina	Heart pain due to coronary artery disease
Anti arrhythmic	A drug that prevents or controls abnormal heart rhythms
Anticoagulant	A substance that inhibits blood clotting, and is used to treat thrombosis
Arrhythmia	An abnormal heart rhythm, either rapid or irregular or both
Atrial fibrillation and flutter	Very rapid beating of the heart's receiving chambers—the atria
Atropine	Used to quicken an abnormally slow heart rate during a heart attack
Chronotropic	Having an effect on the heart rate
Codeine	An opium alkaloid that controls diarrhoea and mild pain
Colchicine	Relieves acute gout
Constrictive pericarditis	A disease where the heart is encased in dense fibrous tissue—which is relieved by removing it (pericardiectomy)
Cyclosporin	A drug used to prevent organ rejection in transplant surgery
Dermatitis	Disease of the skin; photo-dermatitis is brought on by exposure to sunlight
Diuretic	A substance that increases the volume of urine thus

	relieving abnormal fluid retention (oedema, dropsy, q.v.)
Dropsy	An old term referring to swelling of the legs and abdomen owing to heart failure
Eclampsia	A disease of pregnancy with convulsions and high blood pressure
Emetine	A drug formerly used in amoebic dysentery
Ergometrine	Used to contract the uterus after childbirth and to produce spasm of the coronary arteries
Erythrocytosis	Abnormally high production of red blood cells
Glycoside	An organic compound made up of one or more sugars combined with another molecule such as a steroid nucleus
Heart block	Also called atrio-ventricular block; an abnormally slow heart rate owing to disease of the conducting system
Heart failure	Inability of the heart to maintain an adequate circulation, leading to oedema (fluid retention) in the body and the lungs
Hepatic cirrhosis	Hardening of the liver
Hyoscine	Used before operations, as pre-medication, to soothe the patient and reduce bronchial secretions
Hypertension	High blood pressure in the body
Intal	Trade name for sodium cromoglycate (q.v.)
Lignocaine	The most important local anaesthetic agent; also a useful anti-arrhythmic (q.v.) (Class 1B)
Mycosis fungoides	A malignant skin disease
Morphine	A powerful analgesic and narcotic drug
Myocardium	The muscle of the heart
Nifedipine	Used for the treatment of angina (q.v.), hypertension (q.v.) and Raynaud's disease (white fingers)
Oedema	Excess fluid in the tissues owing to fluid retention
Oxytocic	A drug that induces uterine contractions
Papaverine	An opium alkaloid (q.v.) which is a vasodilator (q.v.) drug

Pilocarpine	A remedy for glaucoma (high pressure in the eye)
Pregnancy toxaemia	A condition with high blood pressure and oedema (q.v.), also called pre-eclampsia
Procaine	A local anaesthetic; procainamide is an anti arrhythmic (q.v.) agent (Class 1A)
Prothrombin	A substance involved in normal blood clotting
Pulmonary hypertension	High blood pressure in the arteries of the lungs
Quinidine	A long established anti arrhythmic (q.v.) drug (Class 1A)
Reserpine	A treatment for high blood pressure
Rheumatic fever	Inflammation of the joints and the heart, often a disease of childhood
Stokes-Adams attacks	Attacks of loss of consciousness owing to heart block (q.v.)
Sodium cromoglycate	A drug for asthma with a unique mode of action
Vagal	Pertaining to the vagus nerve, an autonomic cranial nerve supplying the heart, lungs and abdomen
Vasoconstrictor, vasodilator	Substances that either constrict or dilate blood vessels especially the small arteries (arterioles)
Ventricles	The two pumping chambers (right and left) of the heart
Ventricular fibrillation	Very rapid ineffective contractions of the ventricles (q.v.) causing sudden death
Veratrum	Formerly used in eclampsia (q.v.) and in high blood pressure
Verapamil	A class IV anti arrhythmic (q.v.) agent, used also for angina (q.v.) and high blood pressure
Vinca alkaloids	Valuable drugs for the treatment of cancer
Vitiligo	Abnormal white areas on the skin owing to depigmentation

Index

Page numbers in *italics* refer to illustrations

40